Zineb Chaich
Djamel Belatrache
Nadia Saifi

Optimization of Photovoltaic Systems in Arid Zones

Zineb Chaich
Djamel Belatrache
Nadia Saifi

Optimization of Photovoltaic Systems in Arid Zones

Dust impact and cleaning techniques

ScienciaScripts

Imprint

Any brand names and product names mentioned in this book are subject to trademark, brand or patent protection and are trademarks or registered trademarks of their respective holders. The use of brand names, product names, common names, trade names, product descriptions etc. even without a particular marking in this work is in no way to be construed to mean that such names may be regarded as unrestricted in respect of trademark and brand protection legislation and could thus be used by anyone.

Cover image: www.ingimage.com

This book is a translation from the original published under ISBN 978-620-6-72932-7.

Publisher:
Sciencia Scripts
is a trademark of
Dodo Books Indian Ocean Ltd. and OmniScriptum S.R.L publishing group

120 High Road, East Finchley, London, N2 9ED, United Kingdom
Str. Armeneasca 28/1, office 1, Chisinau MD-2012, Republic of Moldova, Europe
Managing Directors: Ieva Konstantinova, Victoria Ursu
info@omniscriptum.com

Printed at: see last page
ISBN: 978-620-3-48782-4

Contents

Energy production represents a major challenge for the years to come, due to the ever-increasing energy needs of both industrialized societies and developing countries. At present, efforts are focused primarily on increasing the use of renewable energy sources, which offer the advantage of being free of charge while helping to mitigate global warming, a phenomenon exacerbated by the increase in greenhouse gas concentrations resulting from the use of conventional energies.

The Algerian government's strategy is based on harnessing inexhaustible resources, with particular emphasis on photovoltaic solar energy. This approach is all the more relevant for Algeria, where sunshine levels are high, reaching an average of 3,000 hours per year over 80% of the national territory. Renewable energy is considered essential to diversify energy sources and prepare Algeria for the future. However, while this strategy is considered realistic, investors are concerned about a lack of visibility in the sector, which is holding back investment in photovoltaics (PV).

To promote the development of this sector, it is crucial to further research and increase the transparency of PV systems, in order to improve their reliability. It is imperative that PV systems operate at their maximum design capacity to ensure a reliable supply of electricity throughout their life cycle. Performance losses due to dust accumulation are a problem that has not been comprehensively addressed, especially in southern Algeria. Consequently, the main objective of this work is to study experimentally the impact of dust on the performance of photovoltaic panels installed in the Ouargla region.

The work is divided into three chapters:

1. The first chapter presents

2. The second chapter is devoted to a literature review on the effects of sand on PV panel efficiency and methods for improving the performance of photovoltaic (PV) systems. This includes a review of previous work on cleaning techniques, providing an in-depth understanding of the approaches adopted to maximize solar panel efficiency.

3.   The third chapter presents the experimental methodology used, as well as the results concerning the impact of dust on the power and temperature of photovoltaic panels. This section highlights the data collected during the tests, analyzing the measurable effects of dust accumulation on system performance.

4.   The present work concludes with a general synthesis, summarizing the results obtained and putting them to use. This conclusion underlines the importance of an ongoing understanding of the factors influencing the efficiency of photovoltaic systems, while identifying avenues for future studies that could contribute to the optimization of these technologies in various contexts.

Chapter ı:

Solar field and photovoltaic panels

I.1. Introduction

Converting solar energy into electricity via photovoltaic cells is a key solution for replacing fossil fuels with sustainable sources. This chapter highlights the characteristics of photovoltaic systems, detailing solar radiation and its various forms in Algeria. We will present the types of solar panels and their advantages, as well as the properties of the semiconductors used. Finally, we look at Algeria's solar energy strategy, its energy transition program, and the solar power plants in operation for photovoltaic electricity generation.

I.2 Solar radiation

Solar radiation consists of solar particles emitted by the Sun and the propagation of electromagnetic waves. Although electromagnetic waves are dispersed in the Earth's atmosphere, a significant amount of energy reaches the Earth's surface. This solar energy takes the form of electromagnetic radiation and other types of radiation, some of which is absorbed by the atmosphere. [1] The types of solar radiation fall into four distinct categories, as shown in figure I.1. The first type is direct radiation, which is solar emission reaching the Earth's surface directly from the sun without being obstructed or scattered in the atmosphere. The arrival of this direct radiation on Earth depends on a number of factors, including the thickness of the atmosphere to be penetrated and the angle of incidence of the sun's rays in relation to the Earth's surface. Measuring instruments such as the pyrheliometer are used to measure the intensity of direct radiation. The pyrheliometer must be equipped with a device that permanently points towards the sun to guarantee accurate measurements of direct radiation intensity [2]. The second type is scattered radiation, which is light dispersed in the atmosphere as a result of its interaction with various atmospheric substances such as small particles, clouds and dust. Scattering means that light is equally distributed in all directions. In the sky, air particles, water droplets (clouds) and dust interact with the sun's rays, scattering them in all directions. The nature and quantity of scattered radiation depend on weather conditions, such as the density of clouds and particles in the air [3].

The third type is reflected radiation or what is also known as reflection, referring to light reflected from the earth's surface or from other objects on its surface. Reflected radiation is influenced by the characteristics of the reflecting surface, such as its texture, color and nature. Reflected radiation can be significant when the earth's surface reflects light significantly, such as highly reflective surfaces like water or snow [4]. The final type is Global radiation is simply the sum of the direct and diffuse components. There are two types of sunshine data: instantaneous radiation, represented by radiation intensity curves as a function of the time of day, and cumulative radiation, which is the sum of global radiation per day. These data are obtained by accumulating all values from year to year and averaging them for each month of the year [5].

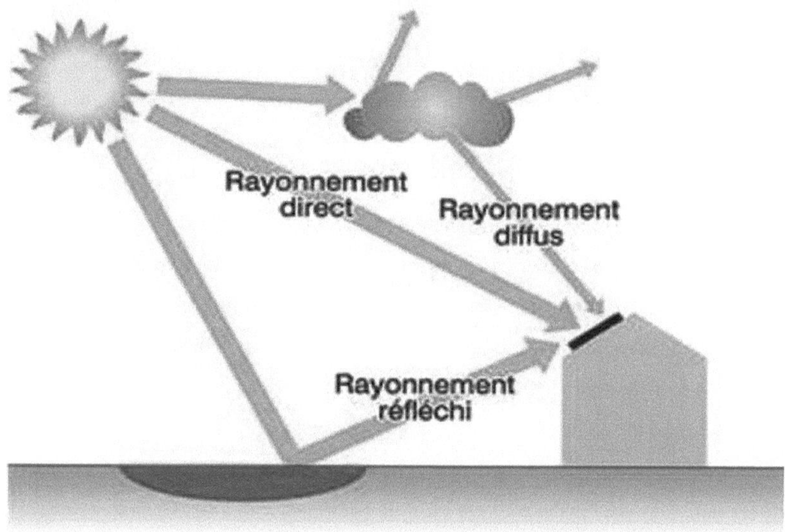

Figure.I.1: Components of solar radiation. [4]

Algeria has considerable solar potential, with remarkable levels of sunshine recorded by the German Space Agency (DLR), reaching up to 1,200 kWh/m²/year in the northern part of the Great Sahara. According to a satellite assessment by the German Space Agency, Algeria stands out as having the highest solar potential in the entire Mediterranean region, estimated at around 169,000 TWh/year for solar thermal and 13.9 TWh/year for solar photovoltaic [6]. Figure I.2 shows the average annual global irradiance received on a horizontal surface over the period 2020.

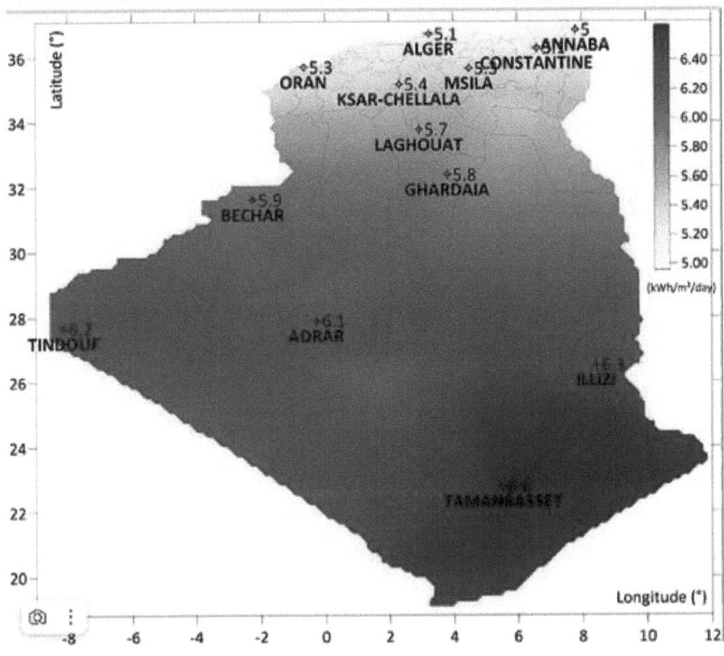

Figure.I.2: Global horizontal irradiance map (2020). [7]

I.3. Photovoltaic solar panels

The solar panel, also known as a photovoltaic generator, is made up of photovoltaic modules interconnected in series and/or parallel to produce the required electrical power. These modules, which capture solar energy and convert it into electricity, are mounted on a metal structure designed to support the solar panel at a specific angle of inclination [8]. There are several types of solar panel used to generate solar energy.

Monocrystalline silicon is currently the most common option for commercial solar cells, despite the availability of many other materials. The term "monocrystalline" means that all the atoms in the active photovoltaic material are part of a simple crystalline structure where there is no disturbance in the ordered arrangements of the atoms. Despite its high cost, it boasts a high efficiency in watts per square meter (around 150 W/m2), saving space where necessary. However, its efficiency decreases under low illumination [9].

Polycrystalline silicon is made up of small grains of crystalline silicon. Solar cells based on polycrystalline silicon are less efficient than those based on monocrystalline silicon. Grain boundaries in polycrystalline silicon impede the flow of electrons and reduce the cell's energy yield. The conversion efficiency of a commercial polycrystalline silicon solar cell ranges from 10 to 14%. It also features a high conversion efficiency of around 100 Wp/m2, and is less expensive than monocrystalline panels [9].

Amorphous silicon (a-Si) is deposited as a thin layer on a glass plate or other flexible support. The irregular arrangement of its atoms gives it poor semi-conductivity. Amorphous cells are used wherever a low-cost solution is required, or where very little electricity is needed, for example to power watches, calculators or emergency lighting. They feature a high absorption coefficient, enabling very thin thicknesses in the micron range. However, their conversion efficiency is low (7-10%), and cells tend to degrade more rapidly under light. They can be integrated on flexible or rigid substrates and are less expensive than other panel types, but they are not the optimal choice due to their low efficiency, which requires larger surfaces to be covered [10].

I.4. Photovoltaic power plants :

Photovoltaic power plants installed as part of Algeria's strategy to exploit renewable energies and use clean energy plus 21 PV power plants. Installed capacity: (344.1 MWp) and energy produced since MES: 865 GWh PV. We will mention a few stations as follows : [9]

Table I.1: The largest number of central PV units in the overall total.

La Centrale	The Power
C.S.PV Ghardaïa	1.1 MWp
C.S.PV BRN Ouargla	10 MWp
C.S.PV Saida	30 MWp
C.S.PV M'sila	20 MWp
C.S.PV Tamanrasset	13 MWp
C.S.PV Souk ahras	15 MWp

C.S.PV Djelfa	53 MWp
C.S.PV Leghouat	60 MWp
C.S.PV Adrar	20 MWp

A/ Ghardaïa PV power plant:

The 1.1 MW photovoltaic power plant on a site some 15 km north of the city of Ghardaïa, near the village of Oued-Nechou. It will have a rated output of around 1100 kWp (kW peak).

The total number of panels is 6089

The total surface area of the site is around ten (10) hectares. Only six (06) hectares are earmarked for the power plant, with the remainder reserved for future expansion.

The projects carried out as part of the experimental phase have been in operation since July 2014, As for the Ghardaia technopole: A team (SKTM/CREDER/CDER) is on site to analyze the output and behavior of the various sub-fields (different panel technologies).

Figure I.3: Ghardaïa photovoltaic power plant.

B/ PV plant BRN Ouargla :

The BRN Sonatrach ENI photovoltaic station is a solar power plant with a capacity of10 megawatts (MW) located in the Bir Rebaa Nord region of Algeria. The solar power plant was commissioned in 2018 and was built by the BRN joint venture, which is equally owned by Sonatrach and ENI; equipped with 31320 photovoltaic solar panels with an estimated area of around 20 hectares, Polycrystalline Panel Type and Maximum Power (Pmax) 325W.The electricity produced is fed into the grid to help run BRN's oil and gas plant.

Figure I.4: BRN Ouargla photovoltaic power plant.

C/ Centrale PV Saida :

Central PV Saida is a photovoltaic power plant located in the commune of Aïn Skhouna, in the wilaya of Saida. The plant has a capacity of 30 megawatts (MW) and was commissioned in September 2015. It is one of the largest photovoltaic power plants in Algeria.

The plant was built by the German company Belectric and financed by the Algerian government. The plant uses crystalline silicon photovoltaic modules to convert sunlight into electricity. The electricity generated by the plant is fed into the national grid.

Figure I.5: Saida photovoltaic power plant.

D/ M'sila PV power plant :

Centrale PV Msila is a photovoltaic power plant located in Msila, Algeria. It has a capacity of 20 megawatts (MW) and inaugurated in December 2017. The plant is owned and operated by Société de Gestion des Stations de Production d'Énergie Renouvelable (SKTM).

Figure I.6: M'sila photovoltaic power plant.

E/ Tamanrasset PV power plant :

Centrale photovoltaïque de Tamanrasset is a photovoltaic power plant located in Tamanrasset, Algeria. It has a capacity of 13 megawatts (MW) and was inaugurated in April 2018. The plant is owned and operated by Société de Gestion des Stations de Production d'Énergie Renouvelable (SKTM).

The Tamanrasset PV plant comprises 4,092 photovoltaic panels installed on a 26-hectare site. The plant generates electricity from sunlight and feeds it into the national grid.

Figure I.7: Tamanrasset photovoltaic power plant.

F/ Souk-Ahras PV power plant :

Centrale PV Souk Ahras is a 15 MW photovoltaic power plant located in the Algerian province of Souk Ahras. It was commissioned in April 2016 and is owned and operated by Compagnie Algérienne d'Électricité et de Gaz (Sonelgaz). The plant consists of 30,000 photovoltaic modules that generate electricity from sunlight. The electricity is then transmitted to the national grid.

Figure I.8: Souk ahras photovoltaic power plant.

G/PV Djelfa :

Centrale PV Djelfa is a photovoltaic power plant located in the commune of Aïn El Ibel in the province of Djelfa, Algeria. It is owned and operated by Société de Gestion des Stations de Production d'Énergie Renouvelable (SKTM). The plant has a total installed capacity of 53 megawatts (MW) and was commissioned in 2016.

The Djelfa PV power plant is an important step in Algeria's journey towards a cleaner, more sustainable energy future. The plant represents a major investment in the country's renewable energy sector, and is expected to create jobs and boost the local economy. It should also help Algeria meet its climate change targets.

Figure I.9: Djelfa photovoltaic power plant.

H/ Laghouat PV plant:

The station has a capacity of 60 Mw, 20MW and 40MW photovoltaic power plants, with polycrystalline silicon modules and fixed supports, 500 KW inverters, 0.4 / 30 kV transformers, 1 MW 2 30 Kv substations.

Services: Basic design and technical specifications of equipment and Construction Design Review on-site technical assistance (resident engineers + short-term expert visits).

Figure I.10: Leghouat photovoltaic power plant.

I/ Adrar PV power plant :

The Adrar photovoltaic power plant is part of the national renewable energy development program set up by the ministry responsible. It is part of the southern production unit of SKTM (Sharikate Kahraba Wa Taket Moutadjadida , an electricity production company) and was commissioned on 12/10/2015. The SKTM photovoltaic power plant covers an area of 40 hectares. It is located 10 km from the city center of the Wilaya of Adrar.

Figure I.11: Adrar photovoltaic power plant.

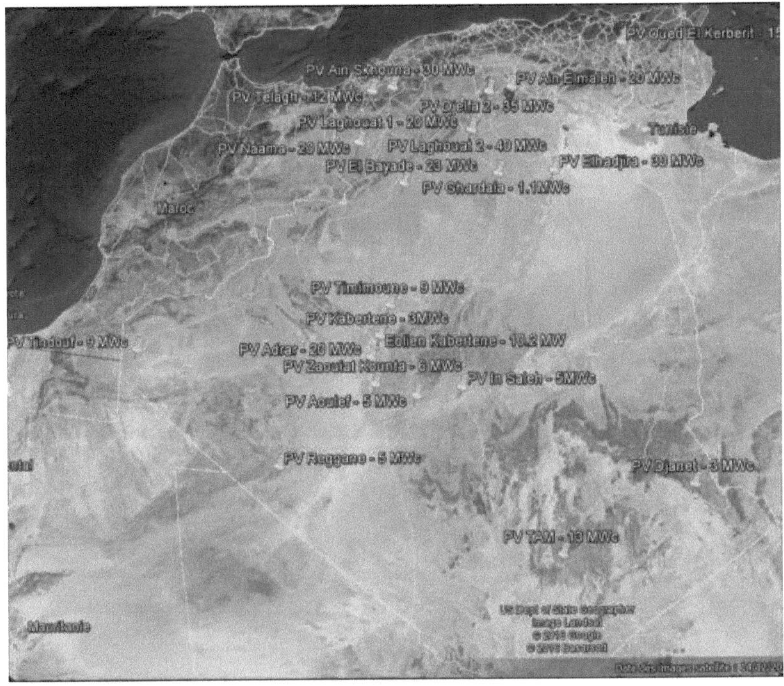

Figure I.12: Photovoltaic power plant project. [14]

I.5. CONCLUSION

This chapter illustrates the Algerian government's commitment to the energy transition, with ambitious programs to transform desert areas into vast photovoltaic solar energy projects. These regions, with their high solar potential and long hours of sunshine, are ideal for this development. In addition, different types of solar panels are analyzed, taking into account their efficiency and economic profitability.

Chapter II:

Effects of Sand on PV Panel Efficiency and Cleaning Solutions

II.1 INTRODUCTION

Solar energy is playing an increasingly crucial role in the global energy transition. The efficiency and operation of photovoltaic (PV) modules depend on various environmental factors, among which dust accumulation is particularly significant. Indeed, dust particles deposited on the surface of PV panels reduce light transmission, directly compromising energy production by hindering the process of converting solar energy into electricity. This chapter explores in depth the impact of dust on PV module performance, and examines the various cleaning methods available to mitigate this undesirable effect. By analyzing these issues, we aim to provide a comprehensive understanding of the challenges associated with the performance of photovoltaic systems, and propose suitable solutions to optimize their efficiency.

II.2 Influence of dust accumulation on the performance of solar photovoltaic (PV) modules.

An urgent problem is the accumulation of dust and contaminants that affects the performance of photovoltaic (PV) panels, posing a substantial risk to their efficiency and operation. Deposited dust prevents the absorption of sunlight, directly affecting the energy conversion process. This accumulation is particularly marked in arid or dusty environments, and is a constant obstacle to increasing the efficiency of solar installations[11-12].

Advanced photovoltaic (PV) technologies focus on developing innovative methods to combat dust accumulation. Researchers are currently exploring new self-cleaning mechanisms to reduce the negative impact of sand on solar panels. [13-14].

Finding cost-effective and efficient techniques to solve this problem is crucial to ensure the longevity, efficiency and durability of solar energy systems, especially in regions vulnerable to dust accumulation. In the face of growing global energy demand, it is essential to improve photovoltaic (PV) technology to address challenges such as dust accumulation. [15-16]

Strengthening the resilience of these systems to minimize inefficiencies consolidates solar energy's role as a reliable and environmentally-friendly energy source, making a significant contribution to the global transition to sustainable energy (Jäger-Waldau [17]).

The Algerian government is focusing on new sources of renewable energy (notably solar power). Since 2011, energy efficiency and renewable energy programs have aimed to produce 22 GW of energy from renewable sources by 2030. However, PV panel performance depends on many factors, including tilt angle, azimuthal orientation, aging, solar radiation spectrum, wind speed at high temperatures, humidity, shadows, snow, mechanical impacts, air pollution, dirt, bird droppings, dust and fouling[18-20].

Many countries are facing the consequences of desertification, with an increase in the frequency of windy days throughout the year.

Many countries are grappling with the consequences of desertification, marked by an increase in the frequency of windy days throughout the year. This increase significantly affects the efficiency of solar energy systems, posing challenges to their proper operation and optimal performance. Previous studies have focused on understanding climatology and the sand cycle, in particular by assessing sand emissions, movement and accumulation.

Optical characterization of dust has been carried out in various regions, including Australia, Africa and Asia. [21-29]

In Iran, the paths of origin and atmospheric factors influencing dust were studied by Salmabadi et al [30], who reported that dust storms in Iran were mainly caused by the Shamal wind. The results showed that over 30% of dust storms in Ahvaz could be associated with the prefrontal model.

Najafpour et al[31]reported that the main sources of dust storms in Tehran were the Euphrates, Tigris and Syrian deserts. Their results confirmed that, on dusty days in Tehran, net solar radiation decreased by around 32-45 W/m^2 due to backscattering by the dust layer.

To identify the factors affecting dust deposition,Lu et al [32] planned three experiments with indoor wind tunnels. The results showed that, in the wind, the density of dust deposition on the solar module decreased with increasing angle of inclination. At 75°, the dust protection efficiency of the super-hydrophobic coating reached a maximum of 94.43%. Han and Lu [33] studied a simplified numerical model to predict the reduction in PV power caused by dust deposition.

In Egypt, Elshazly et al[34] examined the influence of dust deposition on the performance of photovoltaic (PV) panels in the city of El Shorouk. The results indicated that dust deposition reduced performance by 30%. Tilt angle also had an impact on efficiency. The ideal tilt angles for installing PV modules were 15° and 30°, generating the highest energy output.

Fountoukis et al [35] studied the sizes of dust particles deposited on PV panels in Doha, Qatar, and their influence as a function of exposure time. Their study revealed that larger particles caused less degradation than smaller ones, although both types came from the same total amount of dust. This is explained by the fact that smaller particles are more evenly distributed on PV panels, covering a larger surface area than larger particles.

Lasfar et al [36] studied the effect of dust accumulation on PV modules over 50 days in Mauritania. The results revealed that accumulated dust inhibited solar radiation and reduced PV module performance by 21.57%.

The type of dust is crucial. Some researchers have studied artificial dust for laboratory tests. In the UK, Chanchangi et al [37] examined the impact of 13 different natural substances on the efficiency of photovoltaic (PV) panels. Their data showed that ash had the most significant adverse effect on PV module efficiency, reducing it by around 98%, while salt caused the least degradation, at around 7%.

Adıgüzel et al [38] investigated the influence of coal dust of different shapes and masses on the performance of photovoltaic modules under laboratory conditions. Their results revealed maximum power losses of around 62.05% for monocrystalline modules and 60.07% for polycrystalline modules, with a dust mass around 15 g and a coal particle size of the order of 38 μm. In general, the study confirmed that all types of dust negatively affect the performance of photovoltaic panels, with particularly high impacts due to ash, red soil, sand and limestone [39-41].

II.2 Cleaning methods

II.2.1 Manual cleaning

Manual cleaning of photovoltaic panels requires the intervention of an operator using a broom or cloth, supported by suitable structures (Figure II.1). The operator visually assesses the cleanliness of the surface until all dust particles have been removed. This process is laborious and complex, as solar power plants consist of numerous panels installed at heights of 12 to 20 feet or more. This poses problems in terms of time required and safety, both for the operator and the panels. In addition, the use of fluids such as cleaners or gels can affect the transparency of the panels if they are not properly cleaned. The risk of physical damage to panels also remains high and difficult to avoid. [42]

Figure II.1: Manual cleaning of solar panels[42].

II.2.2. Vacuum cleaning

A suction vacuum cleaner is a device which uses an air pump to create a partial vacuum in order to suck up dust and dirt, often from floors, windows, etc. The vacuum cleaner can effectively clean photovoltaic panels, but only on the main surfaces and not in the corners, which requires manual intervention (see Figure II.2). The operator must be properly trained to handle the vacuum cleaner on the panel, as some physical movements are unavoidable. In the long term, dust accumulation on the panel leads to less efficient absorption of solar energy. [42]

Figure II.2: Cleaning with vacuum suction. [42]

II.2.2. Automatic cleaning

5.   Automated cleaning

Automated cleaning with an integrated wiper consists of a rubber blade and a water tank for spraying water with additives and cleaning products. The process is very similar to vehicle window cleaning, and requires an automated mechanism to operate. The device is battery-powered and operates autonomously thanks to an appropriate control system. Although this method works automatically, the results are comparable to those of manual cleaning with similar risks, particularly in terms of efficiency and panel wear potential. [42]

Figure II.3: Automatic dust cleaners [42].

6.   Intelligent cleaning robots

The method involves installing a cleaning robot for each row of photovoltaic panels in a solar power plant, enabling them to be cleaned automatically and regularly without supervision, thus reducing labor costs. The intelligent cleaning robot operates on an independent power supply for cleaning, and has its own energy storage. It uses a waterless cleaning system, which saves energy, protects the environment and conserves water. The frequency of operation can be adjusted as required, and the site can be cleaned regularly to suit site conditions. [42]

Figure II.5: Intelligent cleaning of solar panels. [43]

Siyuan Fan et.al [44] have developed an innovative waterless cleaning robot to remove dust from photovoltaic (PV) arrays in areas where water is scarce. The effectiveness of the waterless cleaning robot was verified by evaluating the light transmission of the panels. The results showed that the average dust removal rate was 92.46%. [44].

I.3.Conclusion

In conclusion, the accumulation of dust on photovoltaic modules represents a major challenge to the optimization of solar energy production. This chapter has highlighted the detrimental effects of dust on PV panel performance and examined various cleaning methods, each with its advantages and disadvantages. While manual techniques can offer targeted cleaning, they carry risks in terms of safety and efficiency. On the other hand, automated solutions offer interesting prospects, but need to be carefully evaluated in terms of cost and environmental impact. As demand for renewable energy increases, it is imperative to continue researching and developing effective cleaning strategies to ensure the long-term performance of photovoltaic systems. By integrating innovative solutions tailored to the specific environments in which these panels are installed, we can maximize their performance and contribute to a more efficient use of renewable energy resources.

Chapter III:

Experimental study and discussion of results

III.1. INTRODUCTION

This chapter focuses on the experimental analysis of photovoltaic module performance, highlighting the impact of dust accumulation on the efficiency of 390 W monocrystalline silicon solar panels. To this end, a comparative study was carried out between two panels: one maintained in an optimally clean condition and the other left in an uncleaned condition. This approach aims to quantify the influence of soiling on the energy production of photovoltaic panels, providing essential data for improving their performance in dust-prone environments.

III.2 Experimental study

III.2.1. Climatic characteristics and geographical location of the Ouargla region

The experiments were carried out under outdoor conditions in the town of Ouargla, located in the northeastern Algerian desert (figure.III.1), at an altitude of 128 meters above sea level, and at the geographical coordinates of longitude 32.1677808 and latitude 4.976654 . [44].

Ouargla's climate is characterized by its dry desert aspect, where drought is manifested in the irregularity and scarcity of rainfall, as well as by permanent dryness, a very wide thermal amplitude, and the presence of a wind regime characterized by hot, dry currents. It also has one of the highest average annual solar radiation values in the world.

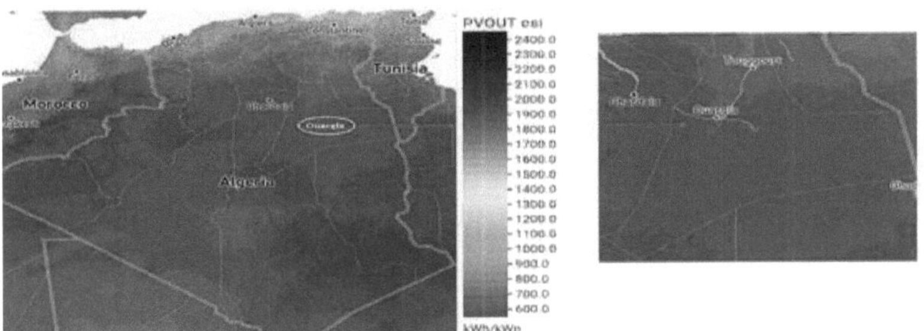

Figure III. III. 1 Average photovoltaic efficiency in Algeria and the Ouargla region over 10 years [45].

Figure III.2 shows the meteorological characteristics of the city of Ouargla from 2013 to 2023. Ouargla is characterized by high levels of sunshine, especially in summer. As for temperature, it reaches values of 50°C in summer, while it does not exceed 22°C in winter. Wind speed is highest in April and May, with an average of around 5.8 m/s. It is lowest in December and January, with an average of around 3.5 m/s. Figure III.2.c shows that Ouargla has a dry climate, with the highest humidity values in winter, at around 50% and 59% for January and December respectively. It then decreases throughout spring and summer, reaching its lowest point of around 16% in July and August. Humidity begins to rise again.

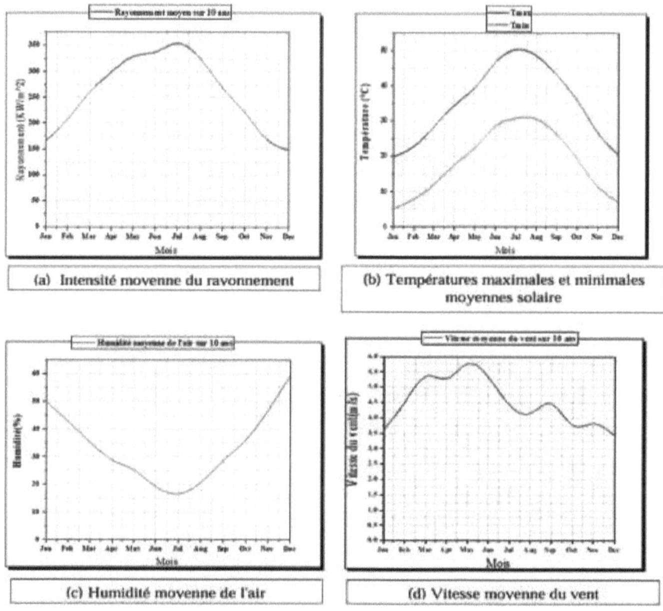

Figure III.2. III.2 Climatic data from 2013 to 2023 in Ouargla [46].

III.2.2. Materials

The materials used in this study are presented below:

7.   Photovoltaic solar panels

The configuration consists of two identical monocrystalline silicon photovoltaic modules (ZGE_FM72_390) with a power output of 390 W (figure.III.4). The modules are tilted at an angle of around 31° and face due south. Specific module parameters are shown in Table.III.1.

Figure.III.3 Photovoltaic module (ZGE_FM72_390).

Table.III.7.PV panel characteristics.

Parameter	Specifications
Maximum power (W)	390 W
Maximum current (A)	9.53 A
Maximum voltage (V)	41 V
Short circuit current (A) (Isc)	10.1 A
Open circuit voltage (Voc)	1. V

8. **Digital multimeter** A GDM-356 digital multimeter was used to measure voltage and current (figure.III.7).

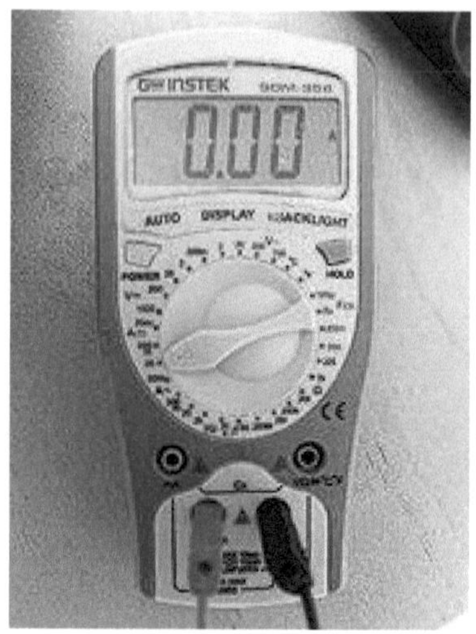

Figure.III.4: Digital multimeter (GW Instek).

9.   Les Résistances

To protect the interface board inputs, two resistors were used (figure.III.8), each with a capacity of 320 W at 600V, 5.7A and 0 to 10 Ω.

Figure.III.5. Resistance.

10.   Solar **radiation meter**

A Hand pyrometer 4890.20 solar radiation meter (Fig.III.6) was used to determine the value of the beam projected onto the solar panel.

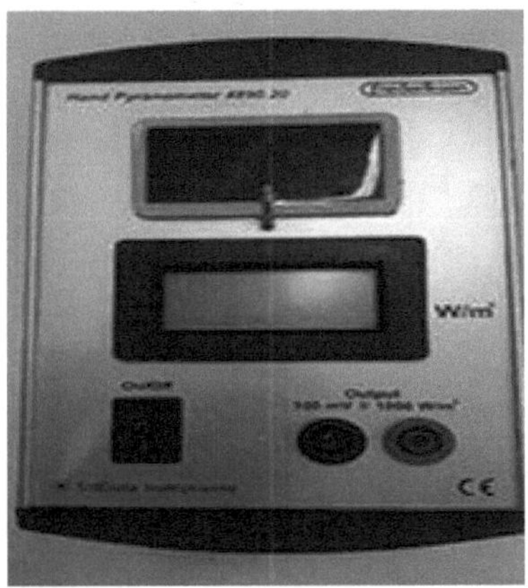

Figure III.6: Hand pyrometer 4890.20 solar radiation meter.

III.2.3. Experimental protocol

To analyze the effect of dust accumulation on photovoltaic panel performance, a comparative study was set up involving two panels:

11. Panel A is designated as the cleaned panel, which will be kept dust-free for accurate comparison.
12. Panel B will serve as an uncleaned panel, allowing a natural accumulation of dust.

This nomenclature guarantees a clear and uniform distinction throughout the experiment. Measured parameters include solar irradiance, voltage and current. Measurements are recorded for each panel every 15 minutes, starting at 08:30 each day. The protocol will be applied over four specific periods starting on 01-05-2022 after 3 days, 7 days, 15 days, 30 days, and 40 days. Each period allows a progressive assessment of the impact of dust over time. At the end of each period, data from each panel will be compared, analyzing performance indicators such as power and efficiency. The study of voltage, current and irradiance measurements will quantify any performance degradation caused by dust on Panel B.

This protocol enables systematic and controlled observation of the effects of dust on PV panels, providing precise information on the implications of dust accumulation over time on performance.

III.3 Results and discussion

In this section, we analyze the impact of dust accumulation on photovoltaic (PV) panel performance by comparing the results obtained for Panel A (cleaned) and Panel B (uncleaned). The parameters measured, such as solar irradiance, voltage and current, enable us to assess the potential degradation in performance of the uncleaned panel over time. Analysis of the data collected over observation periods of 7, 15, 30, and 40 days aims to quantify the cumulative effect of dust on energy efficiency, providing insight into the power losses associated with this accumulation.

The results discussed will enable us to identify the relationship between the duration of exposure to dust and the reduction in panel efficiency, and to understand the mechanisms underlying this degradation. Finally, a comparison of the performance of the two panels will highlight the importance of regular maintenance in dust-prone environments, with a view to optimizing solar energy production.

Figure III.7 shows the variation in power of panels A and B over time. It can be seen that panel A has a slightly higher output than panel B. This difference can be explained by the accumulation of dust on panel B, which reflects part of the solar radiation and thus reduces its efficiency. This difference is explained by the accumulation of dust on panel B, which reflects part of the solar radiation and thus reduces its photovoltaic efficiency. The dust forms an obstructive layer, limiting the optimal absorption of sunlight and reducing the energy conversion performance of panel B.

The variation in power of panels A and B as a function of time, as well as solar irradiance, are illustrated in figures III.8. It shows that the power evolution of both panels A and B follows two distinct phases:

Phase 1 (8 a.m. to 11.30 a.m.): The output of both panels rises significantly with increasing solar irradiance. This increase in power reflects optimum solar intensity captured by both panels.

Phase 2 (11:30 a.m. to 2 p.m.): We observe a fluctuation in the power generated by the two panels, attributable to a drop in solar irradiance caused by cloud cover. This drop in light intensity results in a temporary reduction in panel performance.

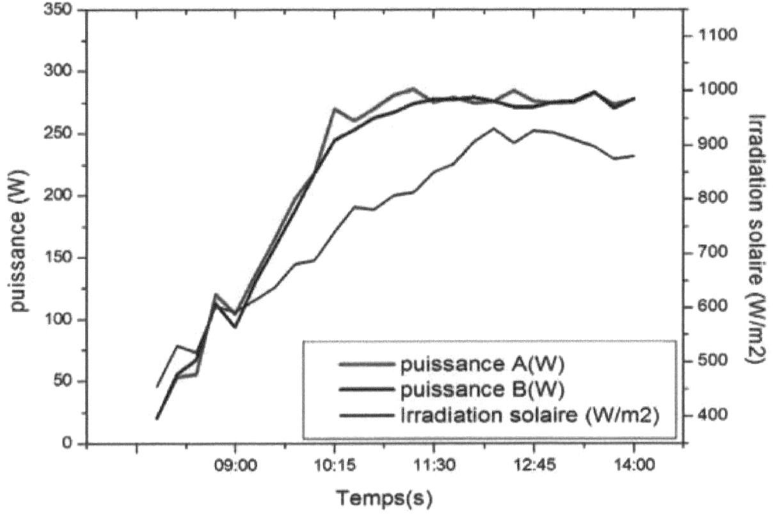

Figure III.7: Variation in solar power and radiation as a function of time after 3 days.

Figure III.9 shows the power evolution of panels A and B as a function of time. Between 8 a.m. and 10:30 a.m., a significant increase in power is observed for both panels, rising from 25 W to 275 W, with a slight difference in favor of panel A (cleaned) compared to panel B (uncleaned). This increase in power is the result of higher solar irradiation. From 10:30 a.m. to 2 p.m., the power of both panels continued to increase, but more moderately, oscillating between 275 W and 305 W, in response to fluctuations in solar irradiance during this period. These observations demonstrate the direct influence of solar irradiation on panel performance, and highlight the reduction in output caused by the accumulation of dust on panel B.

Figure III.10 shows the variation in power of the two panels, A and B, and solar irradiance as a function of time. A notable difference in power is observed between panel A and panel B, the latter having been left without cleaning for over a month.

This difference is explained by the accumulation of dust on panel B, which reduces its efficiency by reducing the amount of sunlight absorbed.

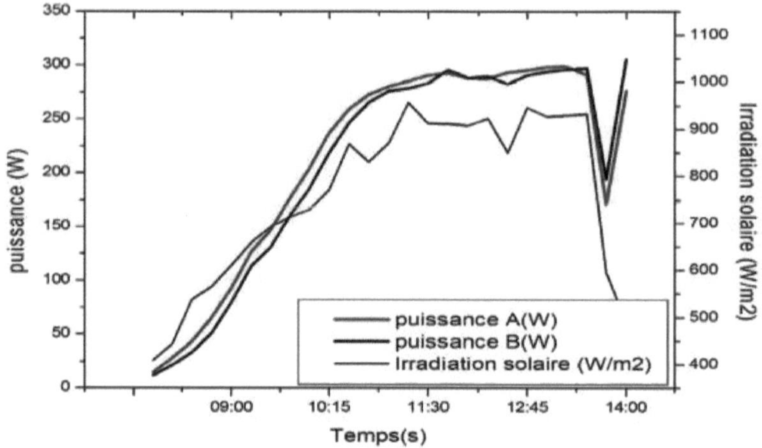

Figure III.8: Variation in solar power and radiation as a function of time after 7 days.

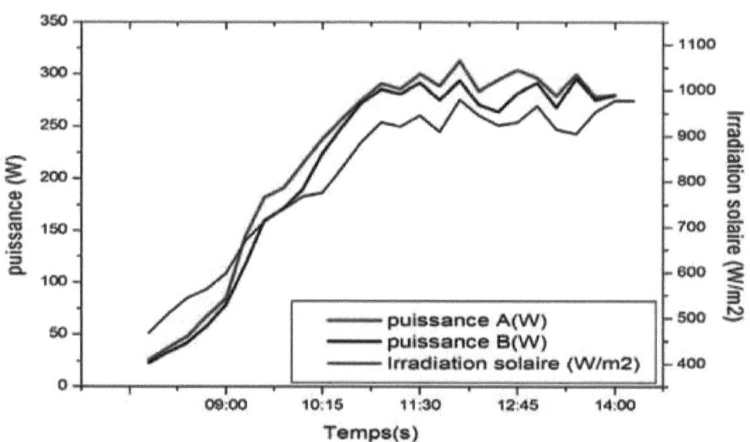

Figure III.9: Variation in solar power and radiation as a function of time after 15 days.

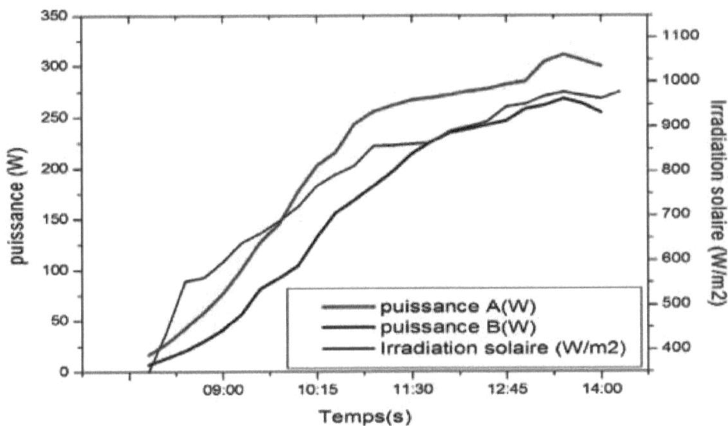

Figure III.10: Variation in solar power and radiation as a function of time after 30 days.

Figures III.11 show the variation in power for panels A and B, as well as solar irradiance, as a function of time. A significant difference is observed between the power values of panel A and panel B after 40 days of exposure, compared with the previous series. This result confirms the impact of dust accumulation on solar panel performance: dust particles accumulated on the surface of panel B reflect a significant proportion of solar radiation, reducing the amount of light absorbed. Consequently, this reduction in available solar radiation directly affects the overall efficiency of photovoltaic systems.

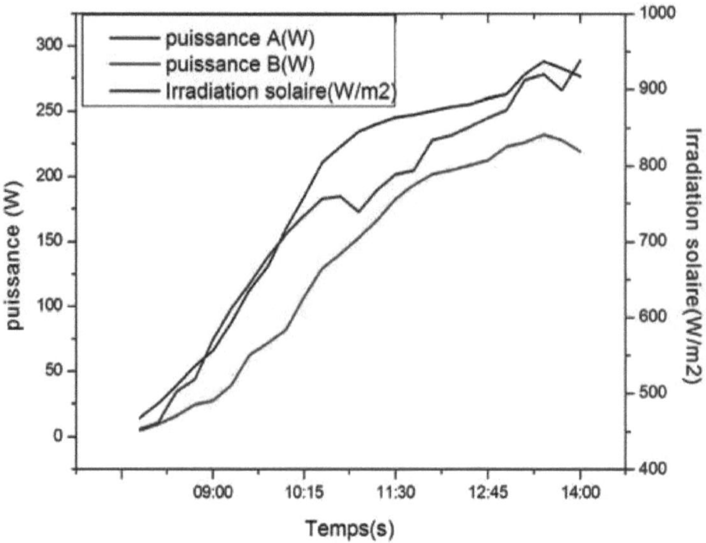

Figure III.11: Variation in solar power and radiation as a function of time after 40 days.

The results show a significant power reduction of 18.68% after 40 days of exposure. This reduction is explained by the fact that dust reduces the solar radiation incident on the photovoltaic module, leading to a deterioration in its performance. As a result, the percentage of energy lost increases in proportion to the accumulation of dust on the panel surface, especially when the panel is exposed to climatic conditions without periodic cleaning. This phenomenon is corroborated by a number of studies which show that dust accumulation on solar panels can lead to significant losses in efficiency, reaching up to 30% in some cases, depending on the nature and thickness of the accumulated particles.

III.4.CONCLUSION

In conclusion, our experimental study revealed that prolonged absence of cleaning has a negative impact on the yield of photovoltaic panels. Regular cleaning of solar panels is therefore essential, with the optimum frequency varying according to the region of installation and climatic conditions. In addition, cleaning should be carried out immediately after dust storms, to minimize particle accumulation and its deleterious effects on performance.

GENERAL CONCLUSION

Demand for electricity continues to grow exponentially, due to the relatively short lifespan of fossil fuels. Against this backdrop, Algeria's renewable energy program promotes the integration of renewable energies through photovoltaic, thermal and wind power. Photovoltaic systems offer many advantages, not least an infinite source of energy, making them particularly well-suited to the electrification of isolated sites and other applications.

However, desert climatic conditions pose challenges, not least the impact of high temperatures and sand winds on the energy efficiency of these systems.

Power loss due to dust accumulation represents a significant challenge for photovoltaic (PV) applications, particularly in arid regions such as the city of Ouargla, characterized by high temperatures, limited precipitation and low humidity. The aim of this study was to experimentally evaluate the impact of dust on the efficiency of photovoltaic modules. Over the course of 40 days, data on solar radiation, voltage and current were recorded manually every 15 minutes between 08:30 and 14:00, comparing two panels: one cleaned before each test and the other left uncleaned.

The results obtained indicate that dust accumulation has a negative effect on the power generated by the photovoltaic module. Panel performance decreases progressively with time of exposure to dust, with respective power drops of 0.039%, 1.41%, 7.37%, 12.45% and 18.68% after 3 days, one week, 15 days, one month and 40 days of exposure.

The results revealed that dust accumulation creates a barrier on the surface of the solar panels, obstructing sunlight and reducing the panels' ability to capture and convert solar energy into electricity. This ultimately leads to a reduction in overall system efficiency, underlining the importance of regular maintenance and cleaning.

To address these challenges, future research should explore advanced cleaning techniques and cooling methods, while integrating predictive models and real-time monitoring systems to improve the performance of photovoltaic installations in dusty environments. Such measures could significantly increase the reliability and efficiency of solar energy systems, particularly in areas prone to heavy dust accumulation.

REFERENCES

[1].https://www.aros-solar.com/fr/le-rayonnement-solaire (accessed March 6, 2024).

[2].URL: https://www.edfenr.com/lexique/rayonnement-direct (accessed March 7, 2024).

[3].URL: https://www.edfenr.com/lexique/rayonnement-diffus (accessed March 9, 2024).

[4].URL: https://energieplus-lesite.be/theories/climat8/ensoleillement-d8 (accessed April 24, 2024).

[5].D.BENATIALLAH, "Détermination du gisement solaire par imagerie satellitaire avec intégration dans un système d'information géographique pour le sud d'Algérie," Doctoral dissertation, Université Ahmed Draia- Adrar, 2019.

[6].Dissertation, Université Ahmed Draia- Adrar, 2019. N. Mebrek, M. T. Bouziane, F. Demnati, and A. E. Nemdil, "Study of the efficiency of a hybrid pumping system (photovoltaic/ electric) for better rural setting management,"

[7].K. Abdeladim, A. Razagui, S. Semaoui, and A. Hadj Arab, "Updating Algerian solar atlas using MEERA-2 data source," Energy Reports, vol. 6, pp. 281-287, 2020/02/01/ 2020, doi: https://doi.org/10.1016/j.egyr.2019.08.057.

[8].T. Mambrini, "Characterization of photovoltaic solar panels in outdoor conditions and according to different technologies Caractérisation de panneaux solaires photovoltaïques en conditions réelles d'implantation et en fonction des différentes technologies," Université Paris Sud - Paris XI, 2014PA112380, 2014. [Online]. Available: https://theses.hal.science/tel-01164783.

[9] B. Flèche and D. Delagnes, "Energie solaire photovoltaïque," STI ELT, June, 2007.

[10].S. Abada and H. Le-Huy, "Study and optimization of a photovoltaic generator for battery charging with a Sepic converter," 2011.

[11] Chanchangi YN, Ghosh A, Sundaram S, Mallick TK (2020) Dust and PV performance in Nigeria: a review. Renew Sustain Energy Rev 121:109704. https://doi.org/10.1016/j.rser.2020.109704 .

[12] He B, Lu H, Zheng C, Wang Y (2023) Characteristics and cleaning methods of dust deposition on solar photovoltaic modules-a review. Energy 263:126083. https://doi.org/10.1016/j.energy.2022.126083.

[13]Shubbak MH (2019) Advances in solar photovoltaics: technology review and patent trends. Renew Sustain Energy Rev 115:109383. https://doi.org/10.1016/j.rser.2019.109383.

[14]Khalid HM, Rafique Z, Muyeen SM et al (2023) Dust accumulation and aggregation on PV panels: an integrated survey on impacts, mathematical models, cleaning mechanisms, and possible sustain able solution. Sol Energy 251:261-285. https:// doi. org/ 10. 1016/j. solener.2023.01.010.

[15]Alami AH, Rabaia MKH, Sayed ET et al (2022) Management of potential challenges of PV technology proliferation. Sustain Energy Technol Assess 51:101942. https:// doi. org/ 10. 1016/j. seta. 2021. 101942.

[16]Allouhi A, Rehman S, Buker MS, Said Z (2023) Recent technical approaches for improving energy efficiency and sustainability of PV and PV-T systems: a comprehensive review. Sustain Energy Technol Assess 56:103026. https:// doi. org/ 10. 1016/j. seta. 2023. 103026.

[17]Jäger-Waldau A (2022) Snapshot of photovoltaics - February 2022. EPJ Photovolt 13:9. https://doi.org/10.1051/epjpv/2022010.

[21]. Schepanski K, Tegen I, Macke A (2012) Comparison of satellite based observations of Saharan dust source areas. Remote Sens Environ 123:90-97. https://doi.org/10.1016/j.rse.2012.03.019

[22]. Alizadeh-Choobari O, Zawar-Reza P, Sturman A (2014) The "wind of 120 days" and dust storm activity over the Sistan Basin. Atmos Res 143:328-341. https:// doi. org/ 10. 1016/j. atmos res. 2014. 02. 001.

[23]. Rashki A, Arjmand M, Kaskaoutis DG (2017) Assessment of dust activity and dust-plume pathways over Jazmurian Basin, southeast Iran. Aeol Res 24:145-160. https://doi.org/10.1016/j.aeolia. 2017.01.002.

[24]. Zhang T, Zang L, Mao F et al (2020) Evaluation of Himawari-8/AHI, MERRA-2, and CAMS aerosol products over China. Remote Sens 12:1684. https://doi.org/10.3390/rs12101684.

[25]. Yousefi R, Wang F, Ge Q, Shaheen A (2020) Long-term aerosol opti cal depth trend over Iran and identification of dominant aerosol types. Sci Total Environ 722:137906. https:// doi. org/ 10. 1016/j. scitotenv.2020.137906.

[26]. Hosseini Dehshiri SS, Firoozabadi B, Afshin H (2022) A new applica tion of multi-criteria decision making in identifying critical dust sources and comparing three common receptor-based models. Sci Total Environ 808:152109. https:// doi. org/ 10. 1016/j. scito tenv.2021.152109.

[27]. Prasad AA, Nishant N, Kay M (2022) Dust cycle and soiling issues affecting solar energy reductions in Australia using multiple data sets. Appl Energy 310:118626. https:// doi. org/ 10. 1016/j. apene rgy. 2022.118626.

[28]. Liu X, Cui L, Tao Q et al (2023) Dust deposition mechanism and output characteristics of solar bifacial PV panels. Environ Sci Pollut Res 30:100937-100949. https:// doi. org/ 10. 1007/ s11356- 023- 29518-1.

[29]. Hosseini A, Mirhosseini M, Dashti R (2023) Modeling of soiling losses on photovoltaic module based on transmittance loss effect. Envi ron Sci Pollut Res 30:107733-107745. https:// doi. org/ 10. 1007/ s11356-023-29901-y.

[30]Salmabadi H, Khalidy R, Saeedi M (2020) Transport routes and poten tial source regions of the Middle Eastern dust over Ahvaz during 2005-2017. Atmos Res 241:104947. https:// doi. org/ 10. 1016/j. atmosres.2020.104947.

[31]Najafpour N, Afshin H, Firoozabadi B (2018) The 20-22 February 2016 mineral dust event in Tehran, Iran: numerical modeling, remote sensing, and in situ measurements. J Geophys Res: Atmos 123:5038-5058. https://doi.org/10.1029/2017JD027593.

[32]Lu H, He B, Zhao W (2023) Experimental study on the super-hydro phobic coating performance for solar photovoltaic modules at Environmental Science and Pollution Research different wind directions. Sol Energy 249:725-733. https:// doi. org/10.1016/j.solener.2022.12.023.

[33]Han Z, Lu H (2023) Numerical simulation of turbulent flow and par ticle deposition in heat transfer channels with concave dimples. Appl Therm Eng 230:120672. https:// doi. org/ 10. 1016/j. applt hermaleng.2023.120672.

[34] Elshazly E, El-Rehim AAA, Kader AA, El-Mahallawi I (2021) Effect of dust and high temperature on photovoltaics performance in the New Capital Area. WSEAS Trans Environ Dev 17:360-370. https://doi.org/10.37394/232015.2021.17.36.

[35] Fountoukis C, Figgis B, Ackermann L, Ayoub MA (2018) Effects of atmospheric dust deposition on solar PV energy production in a desert environment. Sol Energy 164:94-100. https:// doi. org/ 10. 1016/j.solener.2018.02.010.

[36] Lasfar S, Haidara F, Mayouf C et al (2021) Study of the influence of dust deposits on photovoltaic solar panels: case of Nouakchott. Energy Sustain Dev 63:7-15. https://doi.org/10.1016/j.esd.2021. 05.002.

[37] Chanchangi YN, Ghosh A, Sundaram S, Mallick TK (2020) Dust and PV performance in Nigeria: a review. Renew Sustain Energy Rev 121:109704. https://doi.org/10.1016/j.rser.2020.109704.

[38] Adıgüzel E, Özer E, Akgündoğdu A, Ersoy Yılmaz A (2019) Prediction of dust particle size effect on efficiency of photovoltaic modules with ANFIS: an experimental study in Aegean region, Turkey. Sol Energy 177:690-702. https://doi.org/10.1016/j.solener.2018. 12.012.

[39] Javed W, Wubulikasimu Y, Figgis B, Guo B (2017) Characterization of dust accumulated on photovoltaic panels in Doha, Qatar. Sol Energy 142:123-135. https://doi.org/10.1016/j.solener.2016.11.053.

[40] Darwish ZA, Sopian K, Fudholi A (2021) Reduced output of photovoltaic modules due to different types of dust particles. J Clean Prod 280:124317. https:// doi. org/ 10. 1016/j. jclep ro. 2020.124317.

[41] Radonjić I, Pavlović T, Mirjanić D, Pantić L (2021).Investigation of f ly ash soiling effects on solar modules performances. Sol Energy 220:144-151. https://doi.org/10.1016/j.solener.2021.03.046.

[42]. HUDEDMANI, Mallikarjun G., et tout. A comparative study of dust cleaning methods for the solar PV panels. Advanced Journal of Graduate Research, 2017.

[43].Benjamin Figgis, Veronica Bermudez, Juan Lopez Garcia, Effect of cleaning robot's moving shadow on PV string, Solar Energy Volume 254, 2023, Pages 1-7, DOI:10.1016/j.solener.2023.03.003.

[44].Siyuan Fan ,Wenshuo Liang ,Gong Wang ,Yanhui Zhang ,Shengxian Cao , A novel water free cleaning robot for dust removal from distributed photovoltaic (PV) in water-scarce areas. Solar Energy Volume 241, 15 July 2022, Pages 553-563

[44]https://solargis.com/maps-and-gis-data/download/algeria.

[45].https://apps.solargis.com/prospect/map?c=11.523088,8.261719,3&s=26.869814 ,7.77832.

[46] Remund J, Müller S, Schmutz M, Graf P (2020) Meteonorm version 8. METEOTEST. www.meteotest.com

I want morebooks!

Buy your books fast and straightforward online - at one of world's fastest growing online book stores! Environmentally sound due to Print-on-Demand technologies.

Buy your books online at
www.morebooks.shop

Kaufen Sie Ihre Bücher schnell und unkompliziert online – auf einer der am schnellsten wachsenden Buchhandelsplattformen weltweit! Dank Print-On-Demand umwelt- und ressourcenschonend produziert.

Bücher schneller online kaufen
www.morebooks.shop

Printed by Books on Demand GmbH, Norderstedt / Germany